SODA-POP
ROCKETS

Paul Jarvis

SODA-POP ROCKETS

20 SENSATIONAL PROJECTS TO MAKE FROM PLASTIC BOTTLES

CHICAGO
REVIEW
PRESS

First published in the United States of America in 2009 by
Chicago Review Press, Incorporated
814 North Franklin Street
Chicago, Illinois 60610

ISBN 978-1-55652-960-3

Ivy Press
This book was conceived, designed, and produced by
Ivy Press
210 High Street, Lewes
East Sussex BN7 2NS, UK
www.ivy-group.co.uk

Creative Director Peter Bridgewater
Publisher Jason Hook
Conceived by Sophie Collins
Editorial Director Tom Kitch
Art Director Wayne Blades
Designer Lee Suttey
Photographer Andrew Perris
Illustrators Ivan Hissey and Richard Palmer

Printed in China

10 9 8 7 6 5 4 3 2 1

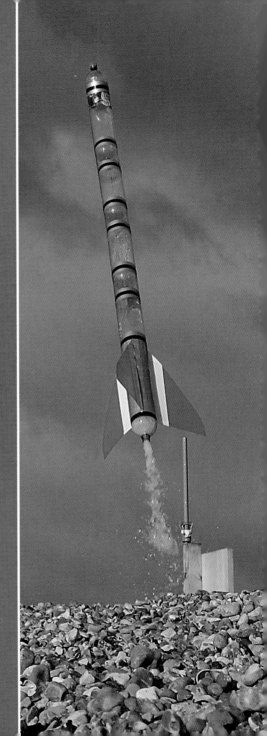

CONTENTS

INTRODUCTION

This book started with the simplest bottle rocket of all, made with an unadorned soda-pop bottle, some water, an air pump, and an improvised bung-and-launch stand. Even this basic model worked reliably and often flew gratifyingly high. But then we started to wonder: What if we stuck two bottles together? Or three? Or five? Added more water? And some fins? What if the rocket had a parachute? What if we used the biggest bottle we could find?

Soda-Pop Rockets is the result of these musings. There are numerous variations and developments you can make on the basic rocket; your designs can be as simple or as elaborate as you have the time and enthusiasm for. We subjected all our rockets to sustained testing at the firing range, and you'll find reports and photographs of the designs in action at the end of each project. Some were sensational; some were consistent; some were unreliable; just a few refused to fly altogether (although you won't find these in the following pages). The what-if question remains, though, however many bottles we launch. You'll recognize fellow enthusiasts by the speculative looks on their faces when they encounter a soda-pop bottle under everyday circumstances.

Some Basic Physics

The physics behind how these rockets work is actually very complex. Reduced to the absolute basics, though, here's what happens when you pressurize and launch a bottle rocket.

When the bottle is partly filled with water and pressurized with air, both the water and the air are under pressure and both want to escape from the bottle. Until launch there is no way out for either. When the rocket is launched, the pressure of the air above the water forces the water out of the bottle neck. The jet of water leaving the bottle causes thrust and the bottle begins to accelerate away from the launch pad. As the water is ejected the bottle becomes lighter and accelerates faster, reaching a speed of about 55 mph (89 km/h) after approximately a quarter of a second. When all the water has been forced out of the bottle, the remaining air will also rush out, adding a little more thrust to the now-empty bottle.

Gravity is trying to stop the rocket from climbing, so the bottle eventually reaches its maximum height (called the apogee). At this point it will start to descend. The descent is slowed by air resistance on the rocket but may still be quite fast, and rockets can break upon landing. The descent can be slowed by incorporating a parachute in the design.

Safety

The projects contained within this book have been designed to be simple, safe, and fun to make and to use. However, there is always a chance of something going wrong, and the author, the publisher, and the bookseller cannot and will not guarantee your safety. Obeying the following safety points will help to minimize the chances of mishaps when you make or launch your rockets, but anything you try from this book you try *at your own risk*.

- The projects are intended to be made and used by adults. Minors should not make them and they should only use them under adult supervision.
- Always use bottles made out of lightweight, ductile plastics, never heavy-duty plastics or glass.
- Always check that your bottle is free from dents, splits, or other damage that may cause it to fail under pressure.
- Never use the rockets in high winds.
- Make sure your launch area is clear.
- Never lean over the rocket when you are preparing it for launch.
- Warn anyone nearby about what you are intending to fire and make sure they stay a safe distance away from the rocket.
- Never aim the rocket in any direction other than straight upward.
- Never aim the rocket at people, buildings, vehicles, power lines, or any other object, whether animate or inanimate.
- Always pressurize and launch your rocket from a safe distance.
- Never overpressurize a rocket.
- Never hold a pressurized rocket.
- It is advisable to wear safety goggles at all times when making and testing the projects.
- Check and follow all local, state, and federal regulations.

PART 1
ROCKET BASICS

This section takes you through a few straightforward rocket designs and shows you how to build two sorts of launcher: one very simple and one sturdy enough to take even the heaviest and most ambitious models from the second and third sections of the book.

Unless you're already a seasoned rocket engineer, try the most basic rocket first. Not only is it easy and quick to make, but it can also achieve surprisingly impressive results at the firing range. Then build your skills and confidence by adding the features that will make your creations look more like real rockets: fins, nose cones, and even a working parachute.

1.1

This starter project is made with a simple improvised launcher that guarantees fast and satisfying results. When you have successfully fired your first rocket, you can try a slightly more elaborate setup using the heavy-duty launcher described in project 1.3 (see pages 22–25).

You will need:

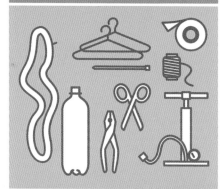

- 2-liter plastic soda-pop bottle
- Duct tape
- Scissors
- Pliers
- Inner tube from a bicycle tire
- Cable tie
- Two wire clothes hangers
- Ball of strong string
- Bicycle pump. A pressure gauge is useful but not essential.

A BASIC ROCKET AND LAUNCHER

1. Cut the valve stem off the inner tube, cutting it as close to the tube as possible. Cut a strip 1 inch (2.5 cm) wide and 24 inches (61 cm) long from the remainder of the inner tube.

2. To make the stopper for the bottle, wrap the strip of rubber from the inner tube tightly around the valve stem until it is the correct size to stop the opening of the soda-pop bottle; it should fit it very snugly. Cut off any excess rubber and secure the rolled stopper with a cable tie. Wrap a layer of duct tape around the stopper. This will improve the seal and make it easier to insert into the bottle opening.

1.

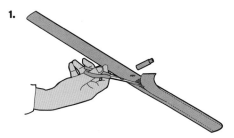

2.

3. Fill the bottle so that it is from one-third to one-half full with water and stop the mouth with the wrapped valve.

4. To make the stand, take a wire clothes hanger and bend it as shown in the diagram. Carefully measure lengths A and B. Length A allows the trigger (see step 5) to fit neatly over the rim of the soda-pop bottle, and length B is the exact distance between the rim of the bottle and the top of the rubber wrapped around the valve. The legs of the stand should about 8 inches (20 cm) long.

5. To make the trigger, straighten out a length of wire approximately 12 inches (30 cm) long from the second clothes hanger and cut it with pliers. Bend it to create a U shape that fits around the neck of the bottle, but under the loops of the stand. Tie a length of string about 13 feet (4 m) long to the center of the trigger's rounded end.

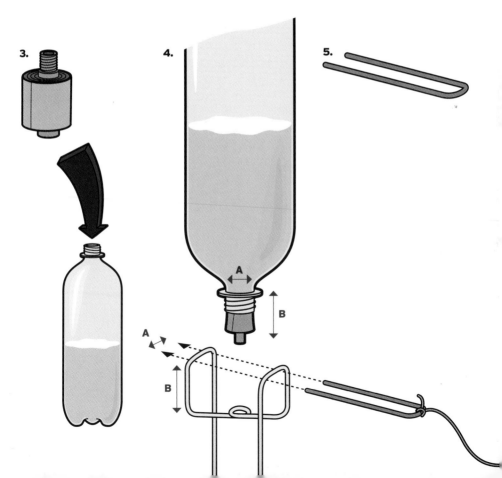

To Set Up the Stand

Push the stand into soft ground. Place the bottle in the stand with the valve pointing downward through the loop in the stand. Hold the bottle steady with one hand, then push the trigger into the stand so that it is wedged between the top of the stand and the rim of the neck of the bottle. This will prevent the rocket from launching prematurely while air is being pumped into it.

To Prepare for Launch

Attach the bicycle pump to the valve and pump air into the rocket. If you have a pressure gauge, a pressure of about 60 psi (pounds per square inch) works well. If you don't have a gauge, you will need to experiment to find the optimum pressure. Keep a record of the number of pumps that were used to pressurize before a flight, and then try repeat flights with more or fewer pumps to find the optimum pressure for each rocket.

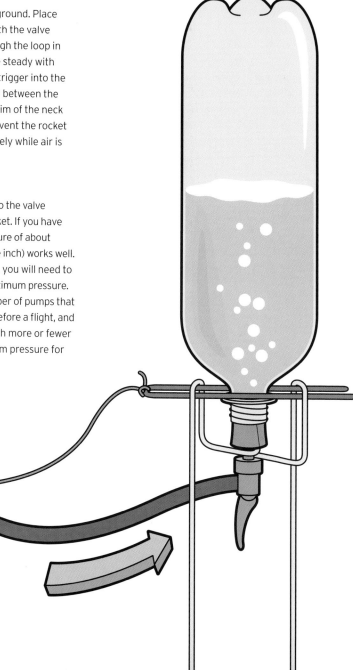

Although you don't need a pressure gauge, it is wise to buy one if you're planning on some sustained rocket experimentation. You'll find being able to keep an accurate record of different pressures invaluable if you want to keep proper flight records.

To Launch Your Rocket

Now that your bottle is pressurized, you are ready to fire. Stand clear and pull sharply the string that is connected to the trigger to release the rocket. When the trigger is released, the pressure inside the bottle forces the stopper out of the neck of the bottle, leaving the stopper behind as your rocket launches into the air.

FIRING REPORT

This rocket was easy and satisfying to fire, but had variable height results, ranging from a mere 10 yards (9 m) up to an estimated 30 yards (27 m). The simplicity of the design makes this an excellent project for a first-time rocketeer.

1.2

Profile

Three sleek fins will not only make your rocket look smarter, they will also help to stabilize it in flight. The model shown here has fins of the classic 1950s-style triangular type, but you can modify the designs according to taste—you'll find another variation to inspire you on pages 44–45. We've used the basic rocket described on pages 10–13 as the foundation for this project. You'll find the fin template on page 108.

You will need:

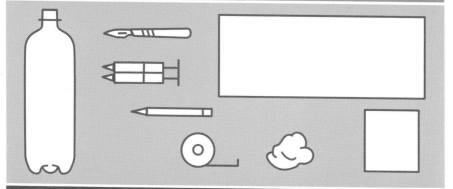

- 2-liter plastic soda-pop bottle
- Epoxy glue
- Heavyweight scissors or craft knife
- Cutting mat
- Soft pencil—2B is ideal
- Tape measure
- Sheet of thin cardstock, 8 x 11 inches (20 x 28 cm)
- Sheet of thin, flexible plastic, 11½ x 16½ inches (29 x 42 cm)— you can find this in hobby or craft stores
- Modeling clay (you'll need a piece the size of a golf ball or a little smaller)

- Either the launcher and the bottle stopper from project 1.1 (see pages 10–13) or the launcher from project 1.3 (see pages 22–25), and a bicycle pump, ideally with a pressure gauge. You will also need access to a photocopier.

A ROCKET WITH FINS

1. In this project you'll be adding fins to a basic bottle rocket like the one you made in project 1.1. The fins must be added before you fill the rocket and insert its stopper.

2. Enlarge the fin template on page 108 to scale onto the thin cardstock, using a photocopier. Cut out the fin shape carefully with scissors or a craft knife. Draw around the cardstock template on the flexible plastic sheet. Repeat twice so that you have the outlines for three fins.

3. Cut around the outlines with scissors or a craft knife. If you are using a craft knife, do your cutting on a cutting mat.

4. Cut the slits on the fins, as shown on the template.

2.

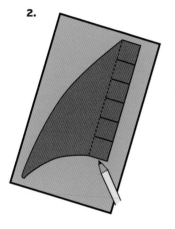

3.

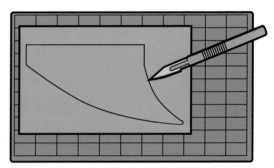

4.

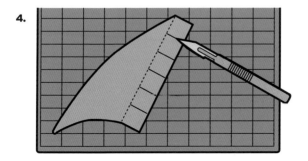

5. Fold three tabs going one way and three going the other on each fin.

6. Make three marks an equal distance apart around the circumference of your bottle, about 2 inches (5 cm) up from the neck where the body of the bottle is straight. Divide the circumference by three and mark it into equal thirds with a pencil.

7. Using epoxy glue, stick the first fin in place. The lowest tab should be stuck about ½ inch (13 mm) above the neck, where the body of the bottle is straight. When the first fin is stuck down, place the other two in position on the pencil marks you made in step 6 and glue them down. Prop up your rocket with books until the glue is dry.

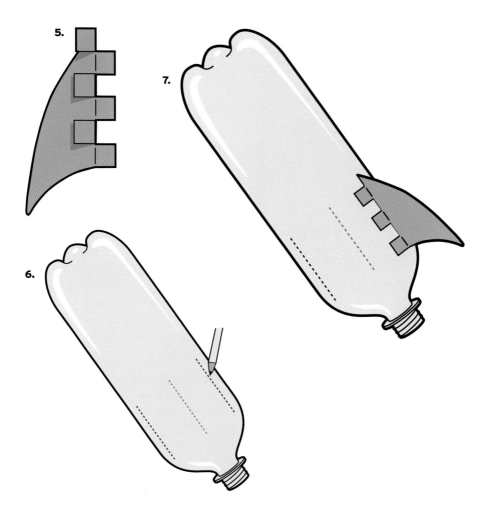

8.

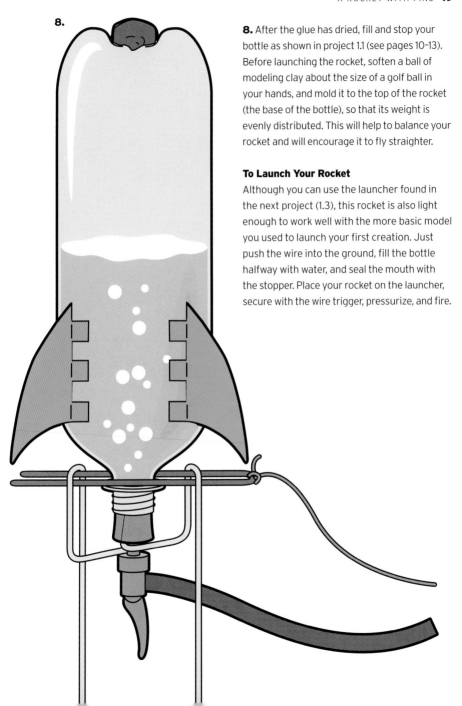

8. After the glue has dried, fill and stop your bottle as shown in project 1.1 (see pages 10–13). Before launching the rocket, soften a ball of modeling clay about the size of a golf ball in your hands, and mold it to the top of the rocket (the base of the bottle), so that its weight is evenly distributed. This will help to balance your rocket and will encourage it to fly straighter.

To Launch Your Rocket

Although you can use the launcher found in the next project (1.3), this rocket is also light enough to work well with the more basic model you used to launch your first creation. Just push the wire into the ground, fill the bottle halfway with water, and seal the mouth with the stopper. Place your rocket on the launcher, secure with the wire trigger, pressurize, and fire.

FIRING REPORT

Fins considerably improved the rocket's stability in flight. We also found that increasing the volume of water and the pressure inside the rocket had marked effects on how high it would go, without affecting its reliability in firing.

1.3

Profile

As you get to grips with larger and more complex rockets, you'll find that you need a launcher worthy of them. This useful, all-purpose launcher is both sturdy and adaptable enough to help any rocket in this book realize its full potential. It's easy to make but will take an hour or two, so set aside an afternoon to put it together.

You will need:

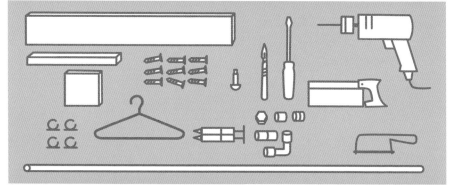

- Wood saw
- Screwdriver
- Small hacksaw
- Craft knife
- Electric drill
- Wood piece, 2 x 2 inches (2.5 x 2.5 cm) square, and 20 inches (0.5 m) long
- Wood plank, 1 inch (2.5 cm) thick, 6 inches (15 cm) wide, and 4 feet (1.2 m) long
- Piece of plywood, ½ inch (13 mm) thick, 8 inches (20 cm) wide, and 12 inches long
- 1¾-inch (4.5-cm) wood screws
- 10-foot (3-m) length of ½-inch (13-mm) OD (outside diameter) plastic water pipe
- ½-inch (13-mm) elbow push-fit pipe fitting

- ½-inch (13-mm) straight-connector push-fit pipe fitting
- Four ½-inch (13-mm) plastic pipe-mounting clips
- ½-inch (13-mm) tank push-fit pipe fitting
- 1-inch (2.5-cm) to ½-inch (13-mm) reducing push-fit pipe fitting
- Wire clothes hanger
- Trigger from project 1.1 (see pages 10-13)
- Two cable ties
- Schrader-type car-tire valve
- 1-inch (2.5-cm) brass compression stop-end pipe fitting
- To test, the rocket from project 1.1 (see pages 10-13) and a bicycle pump, ideally with a pressure gauge

A HEAVY-DUTY ROCKET LAUNCHER

1. Cut the 2 x 2 inch (2.5 x 2.5 cm) wood piece in half to leave two pieces each approximately 10 inches (25 cm) in length.

2. Attach the two pieces to the center of the longer plank in an L shape as shown, using a screwdriver and three wood screws to hold the pieces in place.

3. With the plywood, make a brace to strengthen the L shape, using screws to hold it in place.

4. Using the hacksaw, cut two lengths from the ½-inch (13-mm) water pipe. The first should be 2 inches (5 cm) shorter than the length between the end of the plank of wood and the central upright. The second should be 3 inches (7.5 cm) taller than the height of the upright. Join the lengths of tube together using the elbow fitting.

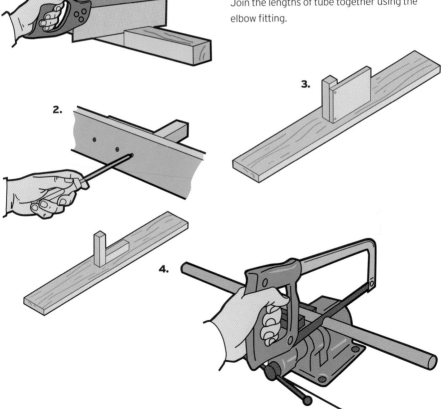

5. Attach the straight connector to the opposite end of the longer piece of tube. Secure the resulting L-shaped piece of piping to the wooden frame using the mounting clips. The shorter piece of tube should be vertically upright against the central upright and the longer piece of tube is fixed horizontally along the plank.

6. Remove the nut from the threaded end of the tank pipe fitting and attach the fitting to the top of the upright tube. Insert a length of ½-inch (13-mm) pipe into the open end of the fitting. (You may need to trim the pipe slightly with a craft knife to make it fit well.) The piece of tube needs to be long enough for the base of the bottle to just clear it when the bottle is in position on the launcher, so use the basic rocket from project 1.1 (see pages 10–13) as a tester as you work.

7. Take the wire clothes hanger and bend it into the frame shape as shown in the illustration. You'll find it easiest to bend the wire accurately using two pairs of pliers. The dimensions of the frame will be determined by the dimensions of the tank fitting you are using. The finished frame needs to hold the bottle tightly against the rubber washer in the tank fitting when the release trigger from project 1.1 (see pages 10–13) is in position.

8. Lower the wire frame down the vertical tube so that the lower loops of the frame are just under the lip under the rubber washer, then tie one of the cable ties around two ends of the lower loops and pull the cable tie tight. Use the second cable tie to pull the opposite ends of the loops together so that the frame is held securely below the lip under the washer.

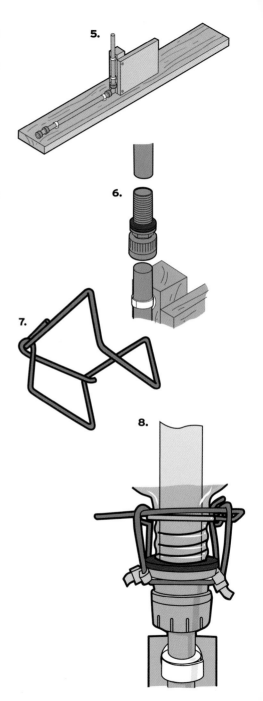

9. Measure the car-tire valve and drill a hole in the brass stop end large enough to fit it tightly. Pull the valve through the hole.

10. Insert the 1-inch (2.5-cm) to ½-inch (13-mm) reducing push-fit pipe fitting into the stop end and tighten the compression fitting, then insert the remainder of the ½-inch (13-mm) pipe into the open end of the reducing push-fit pipe fitting.

11. Push the other end of the pipe into the open end of the launcher assembly. Your rocket launcher is now ready for testing.

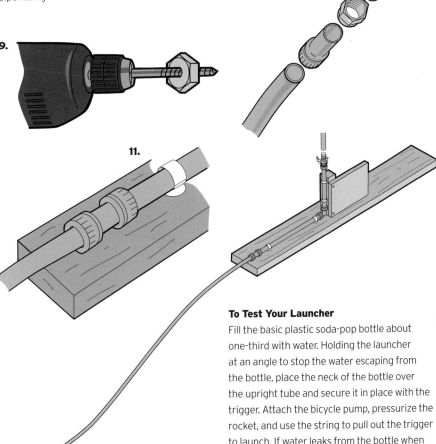

To Test Your Launcher

Fill the basic plastic soda-pop bottle about one-third with water. Holding the launcher at an angle to stop the water escaping from the bottle, place the neck of the bottle over the upright tube and secure it in place with the trigger. Attach the bicycle pump, pressurize the rocket, and use the string to pull out the trigger to launch. If water leaks from the bottle when it's placed on the tank fitting, try adding extra washers on top of the washer on the tank fitting.

FIRING REPORT
Although the launcher was harder to construct than most of the rockets, it served us well for the more demanding and heavier rockets—notably 3.1, the tube-bodied rocket, and 3.6, the large (and very heavy) water-cooler model.

1.4

The two-bottle rocket works on the same principle as the basic rocket described in project 1.1 (see pages 10-13), but is made larger and more impressive by means of a simple duct-tape joint. You can use the bottle stopper and launcher you made for project 1.1 for this rocket, too.

You will need:

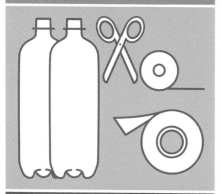

- Two 2-liter plastic soda-pop bottles
- Duct tape
- Heavyweight scissors or craft knife
- Measuring tape
- Either the launcher and the bottle stopper from project 1.1 (see pages 10-13) or the launcher from project 1.3 (see pages 22-25), and a bicycle pump, ideally with a pressure gauge
- If you wish to add fins, follow steps 2-7 of project 1.2 and add them after step 3

A TWO-BOTTLE ROCKET

1. Use the measuring tape to check that the two bottles are the same diameter. For your rocket to fly successfully, the joint between them needs to be tight.

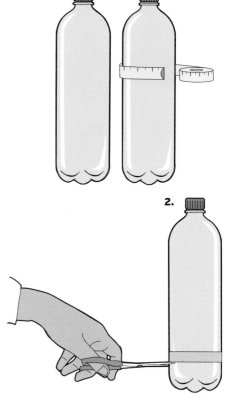

2. Cut a section from the base of one of the bottles using either scissors or a craft knife. Cut as straight a line as possible. If you find it helpful, you can wrap duct tape around the bottle to use as a cutting guide. You can cut as much or as little as you like, but remember that the less you cut, the taller and more impressive your rocket will be.

3. Line up the base of the complete bottle with the cut edge of the shortened bottle and fit the latter carefully over the base of the former. Keep the two parts aligned so your completed rocket will be straight. Cut a length of duct tape a little longer than the circumference of the bottles and tape the cut bottle to the complete one. When you have a neat joint, add another

two layers of duct tape over the first layer to ensure that the joint is firm and won't give way when you launch your rocket.

To Launch Your Rocket

Fill the rocket halfway with water and, if using the launcher from project 1.1 (see pages 10-13), seal the mouth with the stopper. Place your rocket on the launcher, secure with the wire trigger, pressurize, and fire.

3.

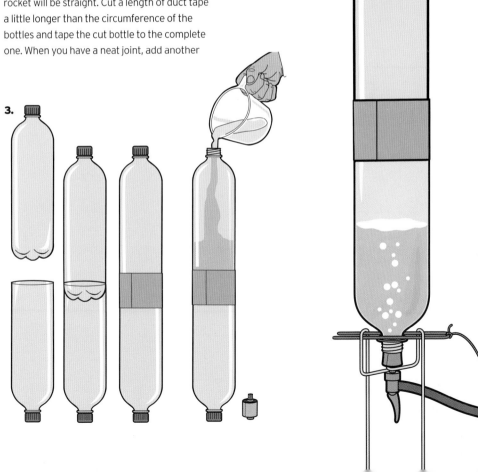

FIRING REPORT
The rocket fired reliably, but with its increased weight it couldn't match the maximum height reached by the basic model in project 1.1. Eventually we amended it and added fins, shown here, to improve its stability.

1.5

Profile

This model, in addition to having fins that will steady its flight on its way up, also has a nose cone containing a parachute that will help it float back down to the ground in elegant style. It's based on a two-bottle rocket like the one described in project 1.4 (see pages 28-29). The top bottle makes a store for the parachute, while the lower bottle holds the water that will propel the rocket's flight.

You will need:

- Two 2-liter plastic soda-pop bottles
- Duct tape
- Epoxy glue
- Heavyweight scissors or craft knife
- Cutting mat
- Hole punch
- Soft pencil—2B is ideal
- Measuring tape
- Large black plastic garbage bag
- Ball of thin string
- Sheet of thin cardstock, 8 x 11 inches (20 x 28 cm)
- Sheet of thin, flexible plastic, 11½ x 16½ inches (29 x 42 cm)—you can find this in hobby or craft stores

- Either the launcher and the bottle stopper from project 1.1 (see pages 10-13) or the launcher from project 1.3 (see pages 22-25), and a bicycle pump, ideally with a pressure gauge. You will also need access to a photocopier.
- If you wish to add fins, follow steps 2-7 of project 1.2 and add them after step 2

A ROCKET WITH NOSE CONE AND PARACHUTE

1. Use heavyweight scissors or a craft knife to cut off the top and bottom of one of your bottles. Don't cut much length off—cut just at the point where the body becomes straight below the neck and above the base, so that you are left with a long cylinder of plastic.

2. Fit one open end of the cut bottle over the base of the second bottle and duct tape the two together neatly. Use two or three layers of tape to ensure that the joint is strong and secure.

3. Make the parachute. Lay the garbage bag out flat on your worktable or on the floor and use scissors to cut off the sealed end, cutting as straight a line as you can. Straighten out the resulting tube and cut down the sides to make two panels of thin plastic. Use one panel to make the parachute and save the other for use in the future.

1.

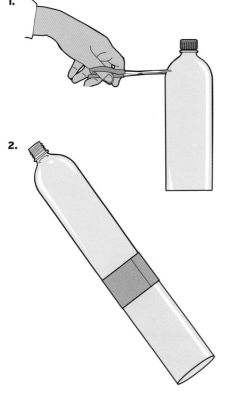

2.

3.

4.

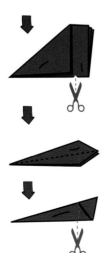

4. Fold the panel in half lengthwise from left to right, then fold in half again from bottom to top. Fold the left edge to meet the bottom edge and make a small cut in the plastic where the top of the left edge meets the bottom edge. Fold the diagonal edge on the left to meet the bottom, then repeat this again to leave a narrow triangle of plastic with the open edges outer-most. Find the small cut you made earlier, then cut in a slightly circular motion to the top edge. Open up the plastic to reveal the large circle.

5. Mark the plastic circle out in quarters (you can do this by eye or use a measuring tape). Cut a 3-inch (7.5-cm) length of duct tape and stick it at one of the quarter-points of the parachute, wrapping it around both sides of the plastic so that it makes a solid tab at the edge. Mark the other three quarter-points with duct tape in the same way, then add three extra tabs between each quarter-point marker, spacing them as evenly as possible. You should now have 16 tabs around the edge of your plastic circle.

5.

6. Use a hole punch to make a hole through the center of each duct-tape tab.

7. Measure the diameter of your parachute with a measuring tape, and cut 16 lengths of string, each 1½ times the length of the diameter measurement. Use a reef knot to tie one end of each piece of string through one of the punched holes in the edge of the parachute.

8. Knot the loose ends of the 16 strings securely together. Cut a short length of duct tape and tape the knotted ends inside the open end of the top of the rocket.

9. Fold the parachute in quarters and place it neatly into the top bottle, pushing the strings in first, followed by the parachute.

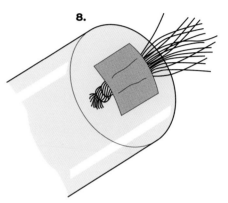

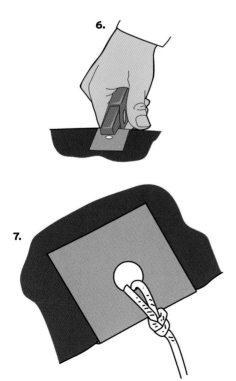

10. Photocopy the pattern for the nose cone of the rocket and its supporting tab to scale onto the sheet of cardstock (see page 108), then cut them out to make a template. Lay the cone template on the plastic sheet and draw around it, then use a craft knife or sharp scissors to cut it out. Draw around the tab three times and cut out the three tab shapes. Fold the tab shapes as shown.

11. Glue the two folded-out edges of each tab at even intervals along the top of the rocket. When glued, the tabs make tiny ledges at a 90-degree angle to the bottle. These will support the cone.

12. Roll the nose cone into shape and tape the tab inside it to hold it together. Cut a length of string 18 inches (46 cm) long and stick one end inside the nose cone with a piece of duct tape. Tie the other end to one of the punched tape ties on the parachute.

To Launch Your Rocket

Fill the rocket with water and, if using the launcher from project 1.1 (see pages 10–13), seal the mouth with the stopper. Place your rocket in the launcher, secure with the wire trigger, and arrange the nose cone carefully on top of the rocket, balancing it on the tabs. Retreat to a safe distance, pump up the pressure, and fire. As the rocket reaches the top of its trajectory, the nose cone will fall off and the parachute will be pulled out of the top of the bottle. The rocket will drift gently back down to earth.

12.

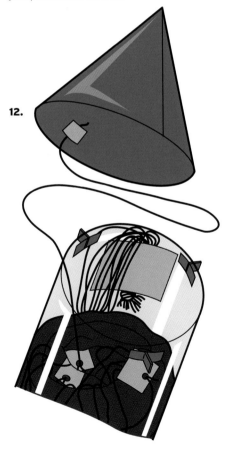

10.

11.

FIRING REPORT
This isn't the most reliable model, but although it did suffer from the occasional malfunction, the sense of satisfaction when the parachute deployed successfully was well worth the frustration of the failed attempts.

1.6

A LINKED-BOTTLE ROCKET

Profile

This pattern also forms the basis for the three- and five-bottle rockets described later in the book. The design performs impressively, surviving multiple firings.

You will need:

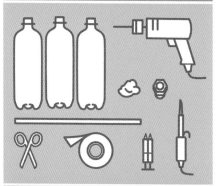

- Three 2-liter plastic soda-pop bottles
- Duct tape
- Epoxy glue
- Heavyweight scissors or craft knife
- Electric drill
- Soldering iron
- Narrow stick or piece of pipe about 3 feet (1 m) long—a wooden plant stake is ideal
- Modeling clay
- ⅜-inch (10-mm) brass compression water-pipe fitting
- Either the launcher and the bottle stopper from project 1.1 (see pages 10-13) or the launcher from project 1.3 (see pages 22-25), and a bicycle pump, ideally with a pressure gauge
- If you wish to add fins, follow steps 2-7 of project 1.2 and add them after step 6

1. Heat the soldering iron and carefully use it to melt a hole in the base of one of the bottles. The hole should be just big enough to fit the screw of the brass compression fitting. If you don't have a soldering iron, you can work the point of a sharp knife through the base of the bottle and then carefully enlarge the hole with small, sharp scissors, but this is less satisfactory as you run the risk of cracking the bottle.

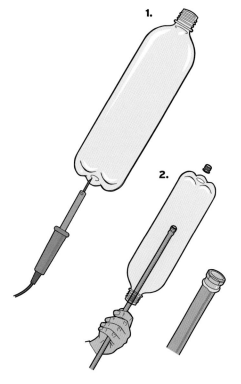

2. Unscrew the brass compression fitting and attach the nut portion to the end of the plant stake using a lump of modeling clay. Push the stake up through the neck of the bottle and screw the other side of the fitting to the nut through the base of the bottle. Add some epoxy glue, pushing it around the screwed-in fitting, to make sure the attachment is firm. Remove the stake and clay through the neck of the bottle.

3. Remove the cap of the second bottle and use the electric drill to make a small hole in it, just large enough to fit the screw of the compression fitting.

4. Fit the cap over the fitting, open end outer-most, and use epoxy glue to stick it to the base of the first bottle.

5. Use scissors or a craft knife to cut a central tube-shaped portion about 8 inches (20 cm) long from the third bottle. Keep the top portion of the bottle and discard the bottom portion.

6. When the epoxy glue is dry, slide the sleeve over the body of the first bottle. Slip the neck of the second bottle into the other end of the sleeve and screw it onto the cap which you previously glued in place. Slip the top of the third bottle over the bottom of the second to form the nose of the rocket. Use two or three layers of duct tape to ensure that all of the joints are strong and secure.

To Launch Your Rocket

Fill the rocket halfway with water and, if using the launcher from project 1.1 (see pages 10–13), seal the mouth with the stopper. Place your rocket on the launcher, secure with the wire trigger, pressurize, and fire.

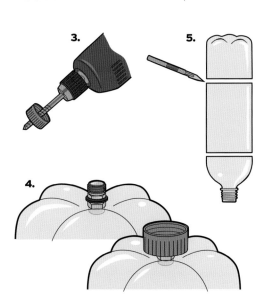

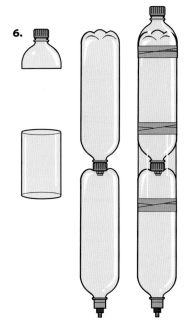

FIRING REPORT
The higher volume of water in a light rocket gave great results at the range. This was a consistent high-flier, although it was prone to crash landing. We tried it with and without fins and it worked well in both incarnations.

PART 2
EXTENDING YOUR REPERTOIRE

Now that you've learned the basic rules, you're ready to explore some new options. Do you want to make different fins, or introduce tubing or ballast to change the shape and heft of your rockets? You'll find out how in the pages that follow. You may even start to think about customizing some of the ideas here to create unique rockets of your own.

Every rocket you make will fly differently, and these pages also offer some exciting high-fliers. The ball-nosed and long-tailed rockets were impressive contenders in both the velocity and altitude stakes.

2.1

A ROCKET WITH CD FINS

Profile

Fins are added to rockets to improve their stability in flight. This project shows you how to add fins made from old CDs to a basic rocket—we chose the linked-bottle rocket from project 1.6, but you can pick any model you like.

You will need:

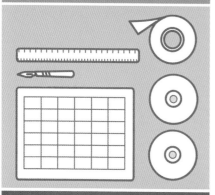

- A linked-bottle rocket as described in project 1.6 (see pages 38-39), or another model rocket of your choice
- Duct tape
- Craft knife
- Cutting mat
- Metal ruler or straight edge
- Two CDs
- Either the launcher and the bottle stopper from project 1.1 (see pages 10-13) or the launcher from project 1.3 (see pages 22-25), and a bicycle pump, ideally with a pressure gauge

1. Place one of the CDs on the cutting mat. Lay the ruler across the CD's center and use the craft knife to score a straight line on the CD.

2. Repeat the scoring three or four times, pressing quite firmly. Then try—gently—to snap the CD into two halves. If it does not snap easily, score it a couple more times. If you force the CD to snap it may not break cleanly. Repeat with the second CD to make four semicircles. You will only be using three for this rocket, so keep the fourth for a future project.

1.

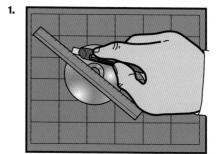

3. Measure the circumference of your bottle rocket and use a pen to mark it into thirds. Holding one of the CD pieces round edge upward, stick a strip of duct tape along either side of the straight edge, then place the fin against one of the lines on the bottle and press firmly on the tape to fix the CD fin in place. Repeat until all three fins are attached to the rocket.

To Launch Your Rocket

Fill the rocket halfway with water and, if using the launcher from project 1.1 (see pages 10-13), seal the mouth with the stopper. Place the rocket in your launcher, secure with the wire trigger, pressurize, and fire.

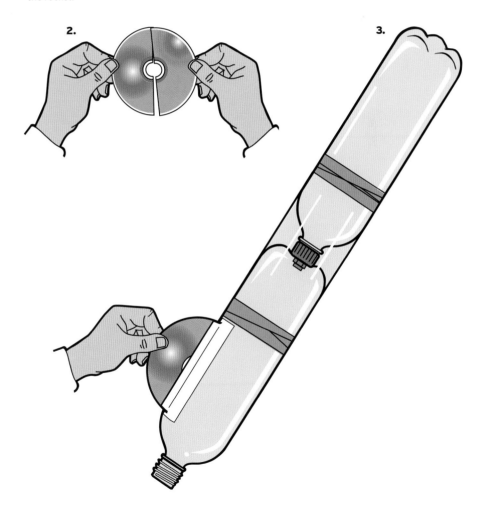

FIRING REPORT

Not the most consistent performer in the repertoire, but the CD fins made for a straight flight when it flew well—with the additional advantage that they stood up robustly to a high number of flights. Less flexible plastic fins would probably have bent or broken.

2.2

Profile

Based on the same format as the linked-bottle rocket in project 1.6 (see pages 38–39), this model uses five bottles—two cut, three whole—to raise the rocket to full three-bottle height.

You will need:

- Five 2-liter plastic soda-pop bottles
- Duct tape
- Epoxy glue
- Heavyweight scissors or craft knife
- Electric drill
- Soldering iron
- Narrow stick or piece of pipe 3 feet (1 m) long—a wooden plant stake is ideal
- Modeling clay
- Two ½-inch (13-mm) brass compression water-pipe fittings
- If you wish to add fins, follow steps 2–7 of project 1.2 and add them after step 6
- The launcher from project 1.3 (see pages 22–25) and a bicycle pump, ideally with a pressure gauge

A THREE-BOTTLE ROCKET

1. Heat the soldering iron and use it to melt holes in the bases of two of the bottles. As in project 1.6 (see pages 38–39), the holes should be just big enough to fit the screw of the brass compression fitting. If you don't have a soldering iron, you can work a sharp knife point through the base of the bottle and then carefully enlarge it with small, sharp scissors. If you choose the latter method, work carefully to avoid cracking the bottle.

1.

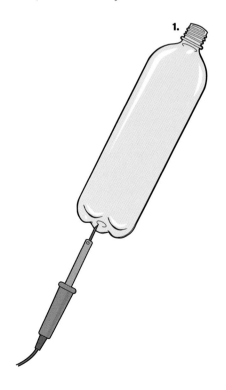

2. Unscrew one of the brass compression fittings and attach the nut to one end of the plant stake or pipe using a lump of modeling clay. Push the stake up through the neck of the bottle and screw the other side of the fitting to the nut through the base of the bottle. Add some epoxy glue, pushing it around the screwed-in fitting to fix the attachment firmly. Remove the stake and clay through the neck of the bottle. Repeat the process with the second compression fitting and bottle.

3. Take two of the bottle caps and carefully drill a hole through each one. Each hole should be just large enough to fit the screw of the compression fitting.

4. Fit the first cap over the fitting, open end outer-most, and use epoxy glue to stick it to the base of the first bottle. Repeat the same process with the second cap and bottle.

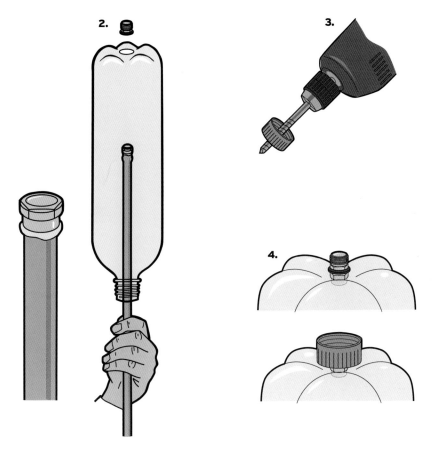

5. Use scissors or a craft knife to cut tube-shaped portions about 8 inches (20 cm) long from the two remaining bottles. These will form the connecting sleeves over the neck joins in your rocket, helping it to keep its shape.

6. After the glue around the compression fittings is dry, assemble your rocket. Slide a sleeve over the body of the first bottle and slip the neck of the second bottle into it. Screw the neck of the second bottle into the cap which you previously glued in place. Secure it firmly in place with bands of duct tape at either end of the sleeve. Repeat the process, sliding a sleeve over the second bottle and fixing the third bottle to the second in the same way.

To Launch Your Rocket

Fill the two lower bottles with water. Place your rocket in the launcher, secure with the wire trigger, pressurize, and fire.

5.

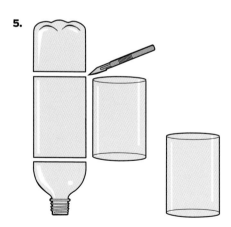

6.

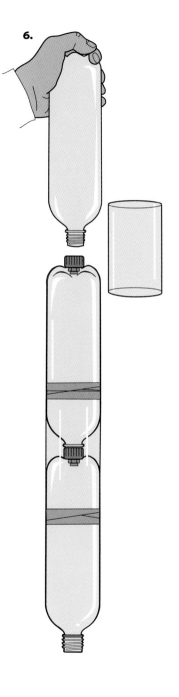

FIRING REPORT
Despite the greater volume of water, this rocket was a rather inconsistent performer. Misfires were relatively frequent, even when fins were added. When it worked, though, it looked great and reached a good height.

2.3

Profile

Fins are integral to the design of this long-tailed rocket and are necessary to steady it in flight. The tubing that forms the tail makes it one of the heavier designs in the book, so it's particularly gratifying that it soars to an impressive height. When you're buying the pipe to make the tail, take an empty plastic soda-pop bottle along with you to double-check that the pipe will fit its neck and can be glued easily in place. If your pipe is too wide for the job, it's almost impossible to produce a neatly formed rocket.

You will need:

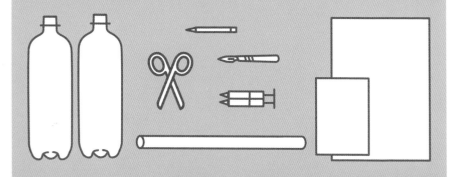

- Two 2-liter plastic soda-pop bottles
- Epoxy glue
- Heavyweight scissors and a craft knife
- 15-inch (38-cm) length of 1-inch (2.5-cm) OD (outside diameter) plastic water pipe—this can be found at building or plumbing supply stores
- Soft drawing pencil—2B is ideal
- Sheet of thin cardstock, 8 x 11 inches (20 x 28 cm)
- Sheet of thin, flexible plastic, 11½ x 16½ inches (29 x 42 cm)—you can find this in hobby and craft stores

- The launcher from project 1.3 (see pages 22-25) and a bicycle pump, ideally with a pressure gauge. You will also need access to a photocopier.

A LONG-TAILED ROCKET

2. Push the pipe into the mouth of one of the soda-pop bottles until the glued length is completely contained in the neck. Leave to dry.

3. Using a craft knife, cut the screw-neck portion away from the second bottle.

1. Take the plastic pipe and smear the outside of one end generously with epoxy glue. Apply the glue in a band around 1 inch (2.5 cm) deep.

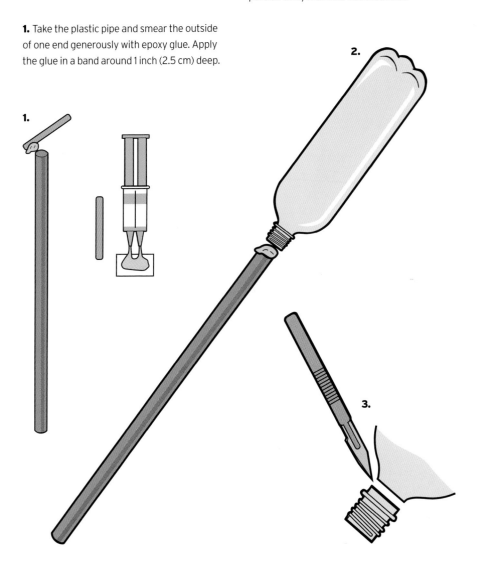

4. Use epoxy glue to stick the free end of the water pipe inside the cut screw-neck piece.

5. Enlarge the fin template on page 109 to scale onto the thin cardstock, using a photocopier. Cut out the fin shape carefully with scissors or a craft knife. Draw around the cardstock template on the flexible plastic sheet. Repeat twice so that you have the outlines for three fins. Cut around the outlines with scissors or a craft knife. If you are using a craft knife, do your cutting on a cutting mat.

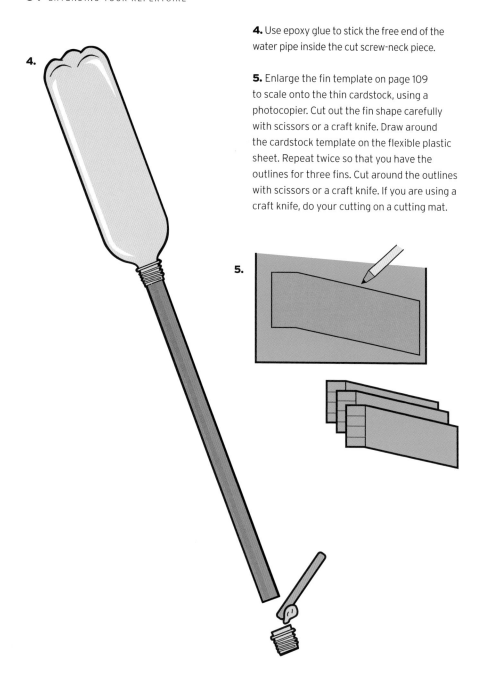

6. Cut the slits on the fins as shown on the template, then fold two tabs going one way and two going the other on each fin.

7. Glue the fins at even intervals around the end of the pipe. They should be glued just above the neck of the bottle.

To Launch Your Rocket

Fill your rocket halfway with water. Place your rocket in the launcher, secure with the wire trigger, pressurize, and fire.

6.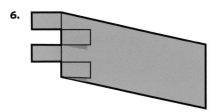

7.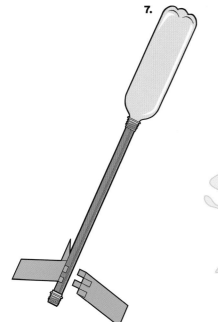

To make your first launch even more impressive, you can fuel your rocket with colored water. Mix a few drops of food coloring into the water before filling the bottle. Green, blue, or red all look good. Watch out for splashes as the rocket takes off!

FIRING REPORT
Completely reliable; a good flight every time, plus a satisfying explosive noise as it left the launcher. It was also a pretty performer in its descent, spinning around gently as it drifted back down to earth.

2.4

Profile

The ball nose stabilizes this rocket and helps to give it additional momentum in flight, and you'll be surprised by how far and how fast it can go. To get the best results, make sure that you choose a really light foam ball. Fins are essential to balance this design—if you make the rocket without them, you'll find that the rocket is prone to turn over in the air well before reaching its full potential.

You will need:

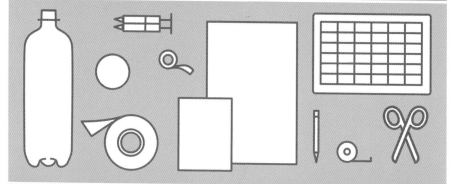

- 2-liter plastic soda-pop bottle
- Duct tape
- Epoxy glue
- Narrow black electrical tape or Scotch tape
- Heavyweight scissors or craft knife
- Cutting mat
- Soft pencil—2B is ideal
- Measuring tape
- Light foam ball—try to match the size to the diameter of the bottle as closely as possible
- Sheet of thin cardstock, 8 x 11 inches (20 x 28 cm)

- Sheet of thin, flexible plastic, 11½ x 16½ inches (29 x 42 cm)—you can find this in hobby or craft stores
- Either the launcher and the bottle stopper from project 1.1 (see pages 10-13) or the launcher from project 1.3 (see pages 22-25), and a bicycle pump, ideally with a pressure gauge. You will also need access to a photocopier.

A BALL-NOSED ROCKET

1. With a craft knife, cut the foam ball in half. You'll only need one half for this project, so save the other for a later rocket.

2. Use duct tape to attach the half ball securely to the base of the bottle. Use several layers of tape. Add a decorative strip of black tape along both edges of the duct-tape joint.

3. Using a photocopier, enlarge the fin template on page 109 to scale onto the thin cardstock. Cut out the fin shape carefully with scissors or a craft knife. Draw around the cardstock template on the flexible plastic sheet. Repeat twice so that you have the outlines for three fins.

1.
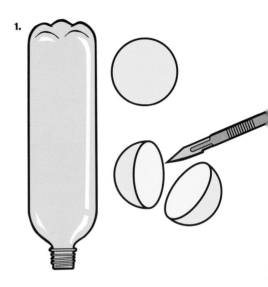

2.
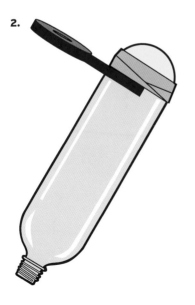

3.
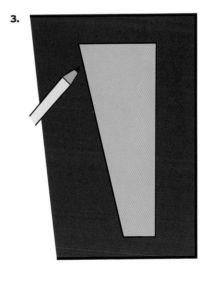

4. Cut around the outlines with scissors or a craft knife. If you are using a craft knife, do your cutting on a cutting mat.

5. Cut the slits on the fins as shown on the template, then fold two tabs going one way and a third going the other on each fin.

6. Stick a strip of the black tape evenly around the body of the bottle just above the neck, where the body becomes straight. Use the top line of the tape as a guide for placing the fins. Measure the circumference of the bottle and mark it in thirds with a pencil. Use epoxy glue to stick the fins in position on the bottle.

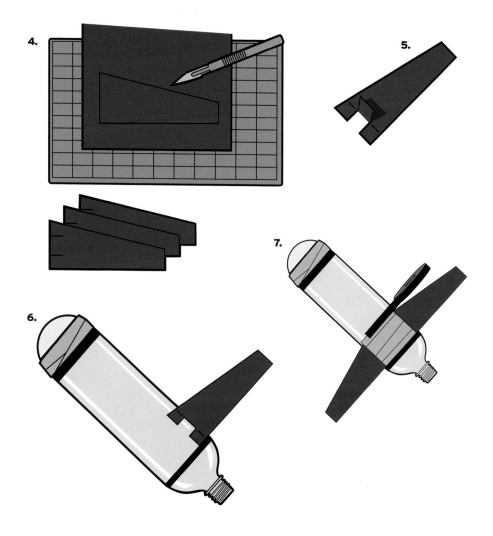

7. Add more duct tape around the fins, using short pieces to fill the space between them and to cover the tabs. Finish off with a decorative strip of black tape along the top and bottom edges of the fins.

To Launch Your Rocket

Fill the bottle halfway with water and, if using the launcher from project 1.1 (see pages 10–13), seal the mouth with the stopper. Place your rocket on the launcher, secure with the wire trigger, pressurize, and fire.

Adding weight to a rocket's nose can have varied results when it is fired. When you've completed this project, try experimenting with different sorts of ballast in your rockets. One of the messiest but most effective of our rockets contained a ripe peach.

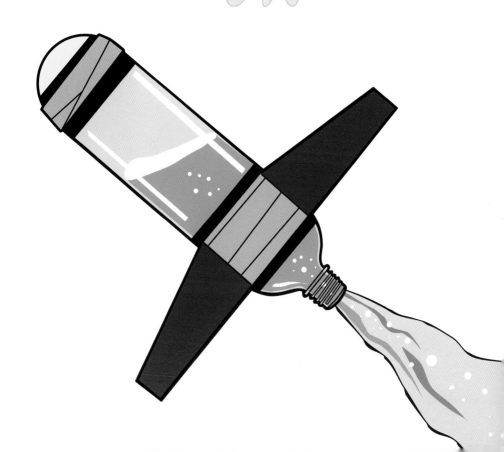

FIRING REPORT

The ball-nosed rocket was a triumph. One of the most reliable models we made, it flew high and very fast. This rocket also fell back to earth more quickly and in a straighter line than most of the other models we built.

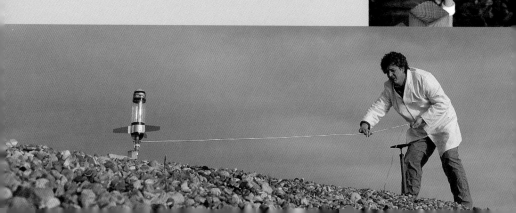

2.5

A ROCKET CAR

If you've ever wondered how far a bottle rocket would go if it were launched horizontally, this design gives you the chance to find out. Make sure you have an open space and a smooth surface to launch onto, and choose a toy car with some weight—this will help it to remain upright as it's propelled along.

1. Make sure that your bottle will rest in a propped, angled position across the back of your toy car. If necessary, experiment with different bottle sizes. Test the bottle stopper with your bottle and add or remove layers of duct tape to ensure it fits tightly.

- 1-liter plastic soda-pop bottle
- Duct tape
- Toy car with open back—a jeep or truck model will work best
- The bottle stopper from project 1.1 (see pages 10-13) and a bicycle pump, ideally with a pressure gauge

1.

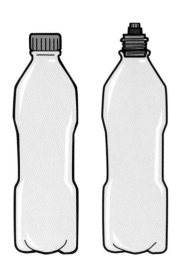

2. Use duct tape to secure the bottle at an angle across the top of the car. The neck of the bottle should extend freely beyond the back end of the car. Keep the bottle opening as low as possible—this will keep the water level above the mouth of the bottle high, resulting in the maximum possible thrust.

To Launch Your Rocket

Fill the bottle halfway with water, seal it with a valve stopper, attach the valve to a bicycle pump, and fill it with air to pressurize. You can't use a launcher with this design, so continue to fill the bottle with air until the pressure forces the stopper out. Make sure there is plenty of space around your car to allow it a free passage.

This very simple project is intended to stimulate your imagination. You can extend the principle to larger items, such as skateboards or even a child's tricycle. Don't forget, though, that the larger the object you want to move, the bigger the soda-pop bottle required to propel it. There are probably some interesting experiments to be made using water-cooler bottles and pedal cars, but we haven't got that far yet.

2.

FIRING REPORT
The rocket car works best with a heavy (or weighted) vehicle; if it's too light, it will leave the ground at the end of the "flight," when the water no longer supplies sufficient ballast to weight it. However, this can look quite impressive.

PART 3
ADVANCED ROCKETS

You've learned the basics and experimented with a few variations. Now it's time to get more ambitious. This final section of the book contains some rockets that are impressive merely by virtue of their size—it's hard to believe that the water-cooler rocket will ever leave the launcher, but it does—and others that will call on all your new-found rocket-making skills. Be very careful when you're rounding the nose of the rocket in project 3.5, and even more so when you ignite the whoosh rocket in project 3.7.

When you've worked your way through these final projects, you're on your own: you'll be a qualified rocketeer able to create blueprints for new rockets that no one has created before.

3.1

Profile

This rocket looks like a missile in flight, due to its elongated shape and sharply triangular fins. You can accentuate the effect by applying a checkerboard pattern to the midsection using squares of black tape. It's one of the heavier rockets, so you need the sturdy launcher from project 1.3 (see pages 22–25), and a large, clear space to launch it—this rocket is far too hefty to risk its making contact with anything on its return to earth.

You will need:

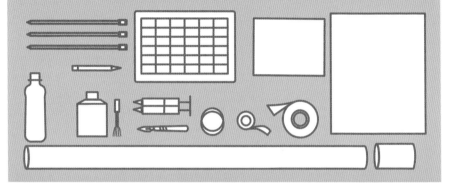

- Small plastic soda-pop bottle with an internal diameter as close as possible to the outer diameter of the ABS tubing (see below)
- Duct tape
- Epoxy glue
- ABS cement—available from building or plumbing suppliers
- Black electrical or Scotch tape
- Craft knife
- Cutting board
- Soft drawing pencil—2B is ideal

- 3-foot (91-cm) length of 2-inch-diameter (5-cm-diameter) ABS tubing—available from building or plumbing suppliers
- End cap and jointing sleeve to fit the ABS tube
- Two or three cable ties
- Sheet of thin cardstock, 8 x 11 inches (20 x 28 cm)
- Sheet of foam board, 11½ x 16½ inches (29 x 42 cm)
- The launcher from project 1.3 (see pages 22-25) and a bicycle pump, ideally with a pressure gauge. You will also need access to a photocopier.

A TUBE-BODIED ROCKET

1. Glue the end cap and jointing sleeve onto one end of the ABS tube using ABS cement. Leave to dry.

2. Cut the small plastic soda-pop bottle in half with a craft knife.

3. Slide the free end of the ABS tube into the upper half of the small bottle.

4. Use epoxy glue to stick the tube into the bottle, and reinforce the joint with cable ties. Secure it further with duct tape.

5. Using a photocopier, enlarge the two fin templates on page 110 to scale onto the thin cardstock. Cut out the fin shapes carefully with scissors or a craft knife. Draw around the cardstock templates on the flexible plastic sheet. Repeat three times with both templates so that you have the outlines for four fins of each size. Cut around the outlines with scissors or a craft knife. If you are using a craft knife, do your cutting on a cutting mat.

6. Measure around the tube and make a pencil mark at each quarter point to place the fins. Use strips of duct tape on either side of each fin to attach them in place on the rocket body. The smaller fins should be attached just above the duct-tape joint at the lower end of the rocket.

7. Make a pencil mark on the body of the rocket 8 inches (20 cm) above the top point of the small fins. Place a tape marker around the tube along the pencil mark and use its top edge as a placement for the lower end of the larger fins. Mark the quarter points with a pencil. Fix the fins with long pieces of duct tape on either side, placing each fin directly above one of the smaller fins.

8. Decorate your rocket with small lengths of black tape. You can use simple stripes or create a checkered pattern on the rocket's body as shown, placing rows of squares at equal intervals in a band around the rocket's body.

To Launch Your Rocket

Fill the tube body halfway with water. Place the rocket on the launcher, secure with the wire trigger, pressurize, and fire.

5.

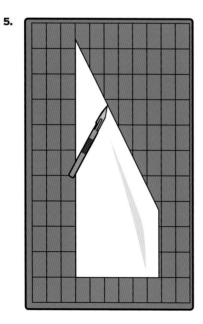

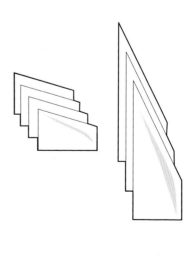

6.

7.

8.

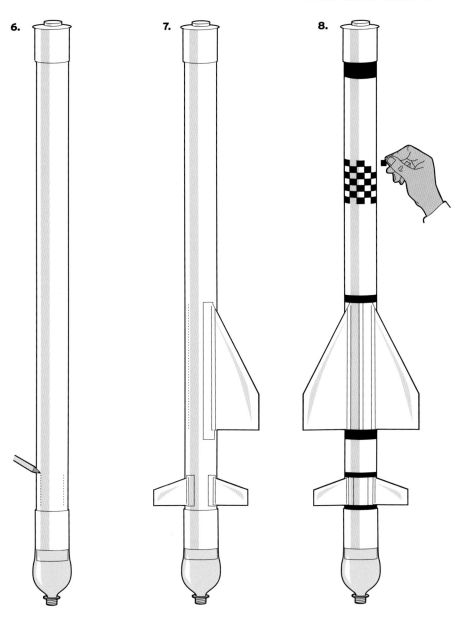

FIRING REPORT
Considering its weight, this model performed reliably. Although it didn't quite scale the heights of some of the other rockets, it never misfired, and its "Cape Canaveral" looks drew favorable comments from passing observers.

3.2

A FIVE-BOTTLE ROCKET

Profile

This is a development of project 1.6. The design connects five bottles in a skyscraper effect.

You will need:

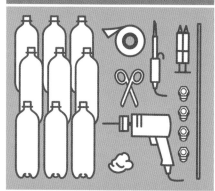

- Nine 2-liter plastic soda-pop bottles
- Duct tape
- Epoxy glue
- Heavyweight scissors or craft knife
- Electric drill
- Soldering iron
- Narrow stick or pipe about 3 feet (1 m) long—a wooden plant stake is ideal
- Modeling clay
- Four ½-inch (13-mm) brass compression water-pipe fittings
- The launcher from project 1.3 (see pages 22–25) and a bicycle pump, ideally with a pressure gauge
- If you wish to add fins, follow steps 2–7 of project 1.2 and add them after step 6

1. Heat the soldering iron and use it carefully to melt holes in the bases of four of the bottles. The holes need to be just big enough to fit the screw of the brass compression fittings. If you don't have a soldering iron, you can work the point of a sharp knife through the bases of the bottles and then carefully enlarge the resulting holes with small, sharp scissors, but this is less satisfactory as you run the risk of cracking the bottles and having to start again.

1.

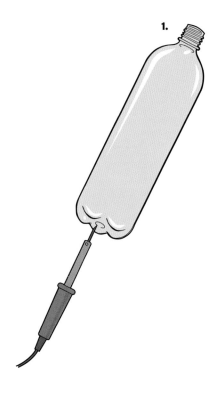

2. Unscrew one of the brass compression fittings and attach the nut portion to the end of the plant stake using a lump of modeling clay. Push the stake up through the neck of the bottle and screw the other side of the fitting to the nut through the base of the bottle. Add some epoxy glue, pushing it around the screwed-in fitting, to make sure the attachment is firm. Remove the stake and clay through the neck of one of the four bottles with holes in. Repeat the process with the other three bottles.

3. Take four of the bottle caps and carefully drill a hole through each one. Each hole should be just large enough to fit the screw of the compression fitting.

4. Fit the cap over the fitting, open end outer-most, and use epoxy resin to glue it to the base of the first bottle.

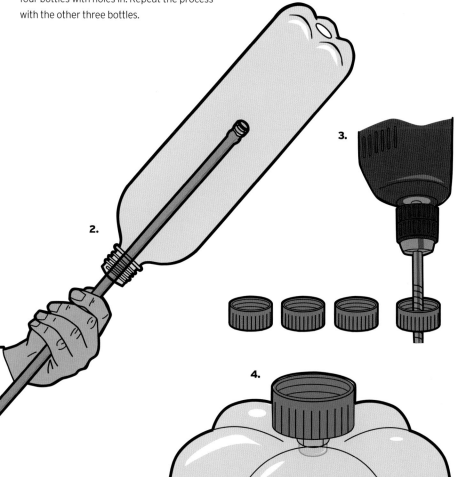

5. Use scissors or a craft knife to cut tube-shaped portions about 8 inches (20 cm) long from four of the remaining bottles. These will form the connecting sleeves over the neck joints in your rocket, helping to hold it in shape.

6. After the glue around the compression fittings is dry, assemble your rocket. Slide a sleeve over the body of the first bottle and slip the neck of the second bottle into it. Screw the neck of the second bottle onto the cap which you previously glued in place. Secure firmly in place with bands of duct tape at either end of the sleeve. Repeat the process, sliding a sleeve over the second bottle and fixing the third bottle to the second in the same way, then repeat again with the third sleeve to fix the fourth bottle to the third. Finally, slide the fourth sleeve over the fourth bottle and fix the bottle that hasn't been used before.

To Launch Your Rocket
Fill the two lower bottles with water. Place your rocket on your launcher, secure with the wire trigger, pressurize, and fire.

6.

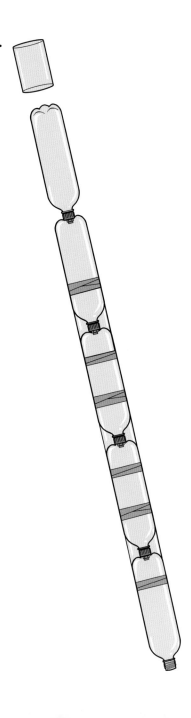

5.

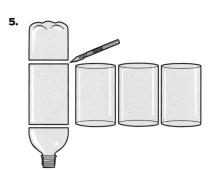

FIRING REPORT

The volume of water helped to sustain this rocket's flight, although its size rendered it too cumbersome to go really high. It invariably launched successfully, but we found that adding fins helped it to climb straight, rather than firing at an angle.

3.3

Profile

In this model, the circular fin steadies the rocket as it fires. This is a small, neat, and simple rocket that flies high and evenly. It's a design that can also benefit from the addition of a nose cone and a parachute if you're feeling more ambitious. Try combining it with the techniques shown in project 1.5 (see pages 32–36) to create a hybrid rocket that both flies with and descends with style.

You will need:

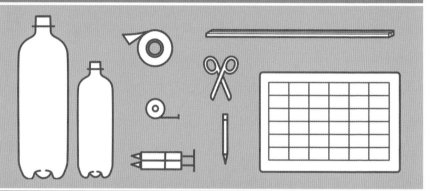

- 2-liter plastic soda-pop bottle
- 1-liter plastic soda-pop bottle
- Duct tape
- Epoxy glue
- Heavyweight scissors or craft knife
- Cutting mat
- Soft drawing pencil—2B is ideal
- Measuring tape
- 3-foot (1-m) length of plastic wire channeling—available from hardware stores

- The launcher and the bottle stopper from project 1.1 (see pages 10–13), with the stopper made to fit the mouth of the 1-liter bottle, and a bicycle pump, ideally with a pressure gauge

A CIRCULAR-FINNED ROCKET

1. To make the circular fin, use a craft knife to cut a section 4 inches (10 cm) deep from the straight body part of the 2-liter bottle.

2. Cut 3 support arms from the plastic channeling, each 5 inches (13 cm) long, using a craft knife and a cutting mat.

3. Measure around the diameter of the 1-liter bottle and mark out thirds with the pencil.

4. Prop the bottle on its neck and stick on the arms at the marked points using epoxy glue. Use small strips of duct tape for added strength, and leave to dry.

5. Use epoxy glue to stick the cut section of the 2-liter bottle to the outer sides of the support arms. Use small strips of duct tape for added strength, and leave to dry. The edge of the fin should be level with the base of the bottle opening to make sure that you can access it for firing.

To Launch Your Rocket

Fill the bottle halfway with water and seal it with the stopper. Place your rocket on the launcher, secure with the wire trigger, pressurize, and fire.

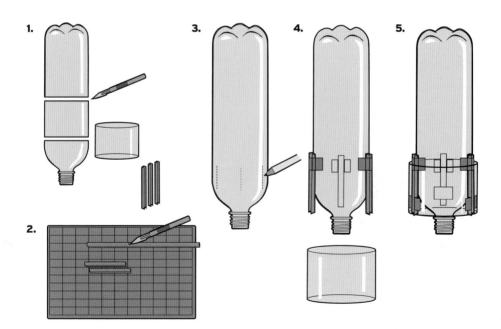

FIRING REPORT
Very fast and high, with a good, straight trajectory. This rocket turned abruptly in the air as soon as all the water had been expelled and came down in a line almost as straight as the ascent. Stand clear of the launch pad until it has returned to the ground!

3.4

Profile

The cluster rocket is a development from project 2.2 (see pages 48–50), the three-bottle rocket, and starts at the point at which that project was completed. The six-bottle addition around its base is adorned with three simple but large fins, giving the cluster rocket a deceptively complicated appearance. Despite its elaborate looks, it's still very easy to make.

You will need:

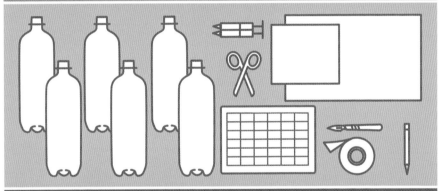

- Three-bottle rocket (see pages 48–50)
- Six 2-liter plastic soda-pop bottles
- Duct tape
- Epoxy glue
- Heavyweight scissors and a craft knife
- Cutting mat
- Soft drawing pencil—2B is ideal
- Sheet of thin cardstock, 11½ x 16½ inches (29 x 42 cm)
- Three sheets of thin, flexible plastic, 11½ x 16½ inches (29 x 42 cm)—you can find this in hobby or craft stores

- The launcher from project 1.3 (see pages 22–25) and a bicycle pump, ideally with a pressure gauge. You will also need access to a photocopier.

A CLUSTER ROCKET

4. Measure the diameter of the rocket and make pencil marks at thirds around the lowest bottle, just above the neck where the bottle body is straight. Glue the fins to the bottle at the marked points using epoxy glue. Add duct tape to strengthen the joints.

1. Make a three-bottle rocket as described in project 2.2 (see pages 48–50).

2. Using a photocopier, enlarge the fin template on page 111 to scale onto the thin cardstock. Cut out the fin shape carefully with scissors or a craft knife. Draw around the cardstock template on the flexible plastic sheet. Repeat twice so that you have the outlines for three fins. Cut around the outlines with scissors or a craft knife. If you are using a craft knife, do your cutting on a cutting mat.

3. Cut the slits on the fins as shown on the template, then fold two tabs going one way and two going the other way on each fin.

3.

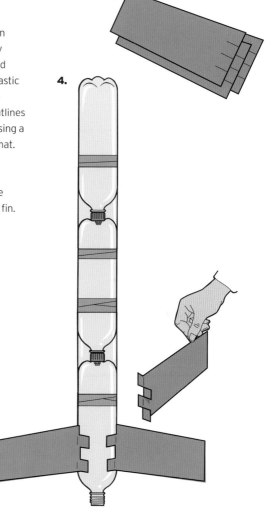

4.

2.

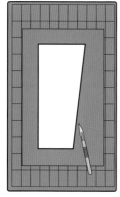

5. Prop your rocket in the firing position, and arrange six bottles around its base, two between each fin. Their bases must be aligned slightly above the top of the neck of the central bottle so they won't interfere with the firing of the rocket. Tape the first bottle in place with duct tape, just above the fin.

6. Continue to tape bottles around the base of the rocket until all six are in place. Run duct tape around the lower parts of the bottles, below the fins, so that the cluster is securely fixed both top and bottom.

To Launch Your Rocket

Fill one and a half bottles with water. Place your rocket on the launcher, secure with the wire trigger, pressurize, and fire.

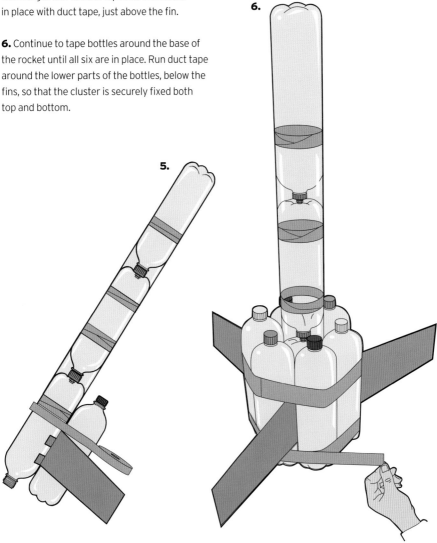

5.

6.

FIRING REPORT
Although this was not one of the most impressive performers in height terms, due to the weight of the extra bottles, it makes a terrific booming sound at the point of takeoff.

3.5

A MOLDED-NOSE ROCKET

Profile

Most of the rockets in this book have noses that are shaped like the bottle bases they are. However, by gently pressurizing your bottle and then warming it, you can achieve a more authentic rounded shape. Here we've applied the technique to the most basic rocket but you can use it on many of the other models, too.

You will need:

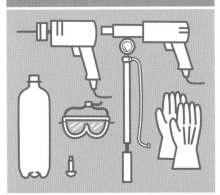

- 2-liter plastic soda-pop bottle
- Electric drill
- Car-tire valve
- Bicycle pump with pressure gauge
- Safety gloves and goggles
- Hot-air paint stripper
- Either the launcher and the bottle stopper from project 1.1 (see pages 10-13) or the launcher from project 1.3 (see pages 22-25)
- If you wish to add fins, follow steps 2-7 of project 1.2 and add them after step 5

1. Drill a hole in the bottle top. It needs to fit the valve tightly, so drill slowly and keep testing the valve against the hole until you have a neat fit.

2. Insert the valve into the bottle top.

3. Screw the cap and valve onto the bottle and, using the gauge and pump, gently pressurize the bottle to 10 psi.

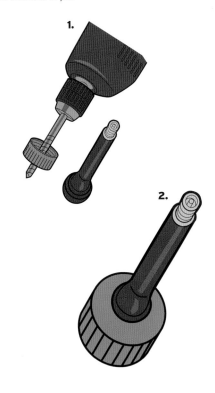

4. Put on the safety goggles and the gloves. Using the hot-air paint stripper, hold the base of the bottle about 10 inches (25 cm) away from the heat and gently warm it. Rotate it as you do so to ensure even heat distribution and to prevent the bottle from overheating. As the plastic warms, it will become flexible and the internal air pressure will cause the heated area to expand, rounding out the end of the bottle.

5. When the bottle base is perfectly rounded, remove the heat source and allow to cool before removing the stopper.

To Launch Your Rocket
Fill the bottle halfway with water and, if using the launcher from project 1.1 (see pages 10–13), seal the mouth with the stopper. Place your rocket on the launcher, secure with the wire trigger, pressurize, and fire.

3.

5.

4.

WARNING
Use extreme caution when heating the bottle. If you hold it too close to the heat it may burst, or, worse, catch fire. Safety goggles and gloves must be worn.

FIRING REPORT
This model looks attractive and its flight is high and fast, although it tends to take off at an angle. If you want a reliably straight flight, the addition of fins—as shown here—will help to stabilize and balance it.

3.6

Profile

This is the big one: the largest rocket you should consider launching in a civilian situation. You may need to befriend your water-cooler supplier to obtain the bottles you need; you can't typically buy them at the store, but we found that most suppliers would happily donate a couple of bottles in return for photos of the rocket in flight. With three gallons of water serving as "fuel," the water-cooler rocket is also very heavy. Take a friend along to the launch—you should have plenty of volunteers!

You will need:

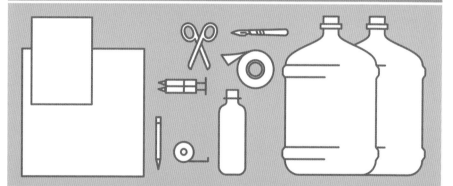

- Two 5-gallon (19-liter) water-cooler bottles
- Small plastic soda-pop bottle—its diameter should fit snugly around the neck of a water-cooler bottle
- Duct tape
- Epoxy glue
- Heavyweight scissors and craft knife
- Soft drawing pencil—2B is ideal
- Tape measure
- Sheet of thin cardstock, 8 x 11 inches (20 x 28 cm)
- Two sheets of thin, flexible plastic, 11½ x 16½ inches (29 x 42 cm)—you can find this in hobby or craft stores
- Either the launcher and the bottle stopper from project 1.1 (see pages 10–13) or the launcher from project 1.3 (see pages 22–25), and a bicycle pump, ideally with a pressure gauge. You will also need access to a photocopier.

A WATER-COOLER ROCKET

1. Cut the top from one of the water-cooler bottles. Choose the point at which you need to cut the bottle—it should be about one-third of the way down—then use a tape measure and a pencil to mark a line around where you want to cut and stick a length of duct tape around the bottle. The upper line of the duct tape will serve as your cutting line. Water-cooler bottles are made from very thick plastic, so use a craft knife and cut slowly and carefully.

2. Fit the partial bottle over the base of the full bottle and attach them firmly together with several rows of duct tape. The partial bottle will form the nose of your rocket.

3. Cut the neck off the small plastic soda-pop bottle using a craft knife, leaving about 2 inches (5 cm) of straight body above the cut.

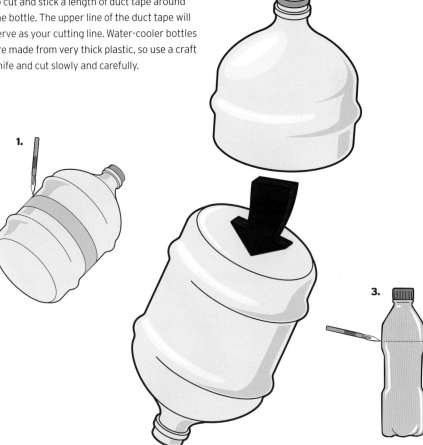

4. Smear a generous layer of epoxy glue inside the cut edge of the small bottle and stick it over the neck of the complete water-cooler bottle. Prop and leave to dry, then strengthen the joint further with several layers of duct tape.

5. Using a photocopier, enlarge the fin template on page 111 to scale onto the thin cardstock. Cut out the fin shape carefully with scissors or a craft knife. Draw around the cardstock template on the flexible plastic sheet. Repeat twice so that you have the outlines for three fins. Cut around the outlines with scissors or a craft knife. If you are using a craft knife, do your cutting on a cutting mat.

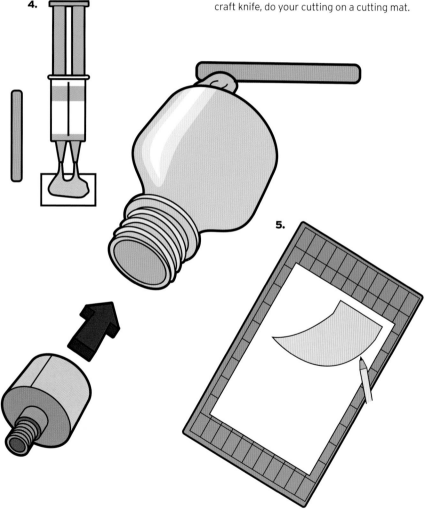

4.

5.

6. Cut the slits on the fins as shown on the template, then fold two tabs going one way and two going the other on each fin.

7. Measure the diameter of the water-cooler rocket and make pencil marks at thirds around it. Glue the fins to the bottle at the marked points using epoxy glue. They should be placed just above the neck of the bottle at the point at which the bottle's body is straight.

To Launch Your Rocket
Fill the rocket with 3 gallons (11 liters) of water and, if using the launcher from project 1.1 (see pages 10–13), seal the mouth with the stopper. Place your rocket on the launcher and ask a helper to steady it, then secure with the wire trigger, pressurize, and fire.

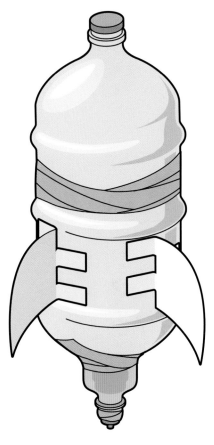

6.

7.

The sheer volume of water in this rocket can give you a thorough soaking if you get too close, so stand back as you launch!

FIRING REPORT

Despite its bulk, this rocket cleared over 50 feet (15 m). From the roar of its launch to the crash as it landed, it was exciting and exhilarating to witness. Stand in the wrong spot and you can get very wet, but as you watch your rocket soar into the sky it will seem well worth it.

3.7

A WHOOSH ROCKET

Profile

For your final rocket, here's something different: the alcohol-fueled whoosh rocket. It is volatile so keep a small fire extinguisher at hand.

You will need:

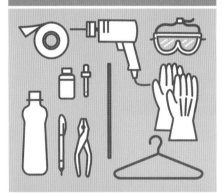

- 2-liter plastic soda-pop bottle
- Duct tape
- Pliers
- Electric drill
- Disposable ballpoint pen
- Wire clothes hanger
- Small amount of isopropanol alcohol—
  this is known as rubbing alcohol and
  can be found at drug stores
- Eyedropper or pipette
- Cable tie
- A long taper (with which to light
  the rocket)
- Matches or lighter
- Safety gloves and goggles
- If you wish to add fins, follow steps 2-7
  of project 1.2 and add them after step 2

1. Unscrew the cap from the soda-pop bottle and carefully drill a hole about ⅜ inch (10 mm) diameter through its center.

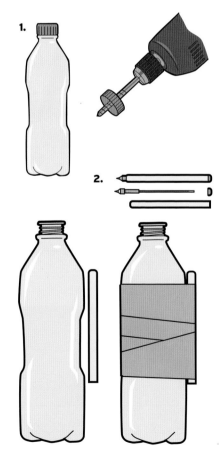

2. Dismantle the ballpoint pen and separate its outer tube from the inner pieces. Duct tape the tube onto the bottle, as shown in the diagram.

3. Use the pliers to bend the coat hanger into the shape of a launcher, as shown. The segment sticking up at an angle (labeled below) should be kept absolutely straight. Attach a cable tie onto the straight segment, leaving enough room for the tube of the ballpoint pen to slide on.

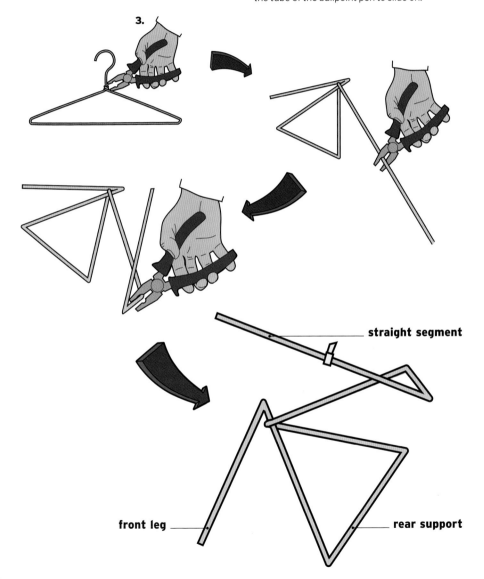

3.

straight segment

front leg

rear support

4. From this step onward, take the rocket outside to an open space. Put on the gloves and goggles. Using the eyedropper, carefully squeeze four drops of the alcohol into the bottle. Replace the cap on the bottle and shake to mix the vapor with the air.

5. Slide the ballpoint-pen tube onto the straight segment of the launcher so that it rests on the cable tie. Place the launcher on a flat piece of ground far away from any objects or people.

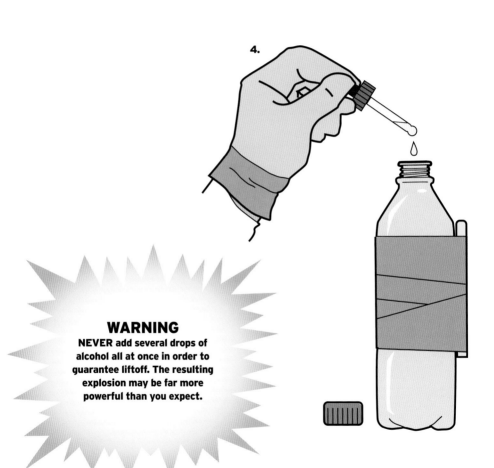

4.

WARNING
NEVER add several drops of alcohol all at once in order to guarantee liftoff. The resulting explosion may be far more powerful than you expect.

To Launch Your Rocket

Light the taper and hold the flame to the vapor that will be emerging from the hole in the bottle cap. The rapid combustion of the alcohol vapor will cause increased pressure inside the bottle, and this pressure will vent through the hole in the cap, causing lift. As the pressure releases, a distinct *whoosh* sound will be heard during liftoff.

WARNING
Be very careful during this launch. It is possible to melt either the cap of the bottle or the bottle itself.

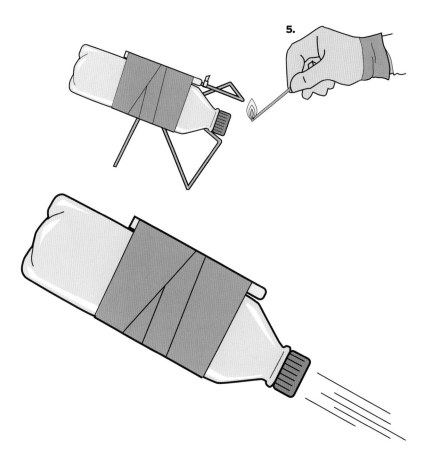

5.

FIRING REPORT
The *whoosh* is just as satisfying as you might expect, and the flight, though brief, is very fast. Any structural addition in the form of fins will be decorative only—with this rocket, blink and you miss it. You'll find, though, that it's well worth repeated flights.

3.8

MAKING A CLINOMETER

The simplest way to measure how high your rockets fly is to make a clinometer. Based on the protractor that is used by geometry students everywhere, a clinometer measures the angle of the rocket at the top of its trajectory. You can then make a simple calculation to work out how high your masterpiece flew.

You will need:

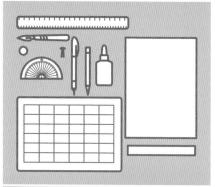

- Strong craft glue
- Craft knife or scissors
- Cutting mat
- Ruler
- Plastic protractor
- Pencil
- Felt-tip pen
- Sheet of medium-weight cardstock, measuring 18 x 12 inches (46 x 30 cm)
- Plastic bottle top
- Metal paper fastener

1. Center the protractor on one side of the sheet of cardstock, aligning it with the edge.

2. Using a pencil, mark 10-degree angles around the edge of the protractor, from 0 to 180 degrees.

3. Use a ruler and a felt-tip pen to extend the lines to the edges of the cardstock at the sides, and to the same length at the top. You should finish up with a large protractor shape, marked with degrees, at one end of the sheet.

1.

2.

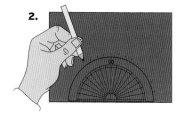

4. Cut a 1½-inch (3.8-cm) strip of cardstock from the other end of the sheet. Cut off the two corners from one end to make a pointed arrow.

5. Punch one hole through the lower end of the arrow, one through the lower edge of the clinometer (1 inch (2.5 cm) in from the lower edge), and one in the center of the clinometer scale.

6. Insert the prongs of the paper fastener through both the clinometer and the pointed arrow. Separate the prongs to secure, but not so tight as to keep the arrow from turning.

7. Glue the bottle top just below the arrow end of the pointer—this will weight the arrow and help it to fall into position.

Your clinometer is now ready for use. Turn the page to learn how.

3.

4.

5.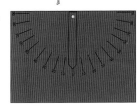

6.

7.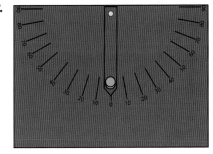

3.9

MEASURING YOUR ROCKET'S ALTITUDE AND TRAJECTORY

Here's how to measure the elevation of your rocket. (Warning: It involves a small amount of mathematical calculation.) You'll need your clinometer and one or more friends to help at the launch.

1. When you're ready to launch, ask your friend to release the rocket. Before he or she does so, measure a distance 10 yards (9 m) away from the launch site and stand at this point, facing the rocket and launcher. Practice holding the the clinometer up to your eye and looking along the top edge so that you will be able to follow the rocket's trajectory as it rises.

2. As the rocket is launched, hold the edge of the clinometer up to your eye and follow the rocket's flight up to its apogee (highest point) as you look along the clinometer's top edge. When the apogee is reached, clamp the pointer in position with your hand. Note the angle of elevation (**E** on the diagram on the next page) as indicated by the clamped pointer.

3. Look at the chart opposite the diagram. To work out the elevation you'll need to apply some trigonometry (you were warned!). A calculator with trigonometry keys will be useful or you can use the chart opposite.

You'll see that there is a triangle between your eye, the apogee of the rocket's flight, and a point directly above the launcher at your eye level.

In this triangle, you know that the distance **D** is 10 yards (9 m), because you measured it earlier. You also know the angle of elevation (**E**), as you have just measured it with the clinometer. In the following formula we will assume **E** to be 40 degrees (°).

To Do the Math:
Height **H** = Distance **D** x tangent (Tan) of the angle of elevation **E**

We can calculate the height:

$$\mathbf{H} = 10 \text{ x Tan } 40°$$
$$= 10 \text{ x } 0.84$$
$$= \text{approximately 8.4 yards (7.5 m)}$$

Because **H** has been measured above your eye level, you also need to add your height up to eye level (**L**) to **H** in order to find the total height of the apogee of your rocket.

This is a reasonably accurate method, but if you want to improve on it you can place two or three other friends, each with a clinometer, around the launch site and take an average of the different results to increase the calculation's accuracy. If you use just one clinometer, your result will only be accurate for a rocket fired straight up. However, most rockets rise at an angle, so by taking more readings from points around the rocket's launch and calculating the average you will achieve greater accuracy.

Elevation (E)	Tan	Height (H) *
10°	0.18	1.8 yards (1.6 m)
15°	0.27	2.7 yards (2.5 m)
20°	0.36	3.6 yards (3.3 m)
25°	0.47	4.7 yards (4.3 m)
30°	0.58	5.8 yards (5.3 m)
35°	0.70	7.0 yards (6.4 m)
40°	0.84	8.4 yards (7.6 m)
45°	1.00	10.0 yards (9.1 m)
50°	1.19	11.9 yards (10.8 m)
55°	1.43	14.3 yards (13.0 m)
60°	1.73	17.3 yards (15.7 m)
65°	2.14	21.4 yards (19.5 m)
70°	2.75	27.5 yards (25.0 m)
75°	3.73	37.3 yards (33.9 m)
80°	5.67	56.7 yards (51.1 m)
85°	11.43	114.3 yards (103.9 m)

*Height (H) if the Distance (D) is 10 yards (9 meters)

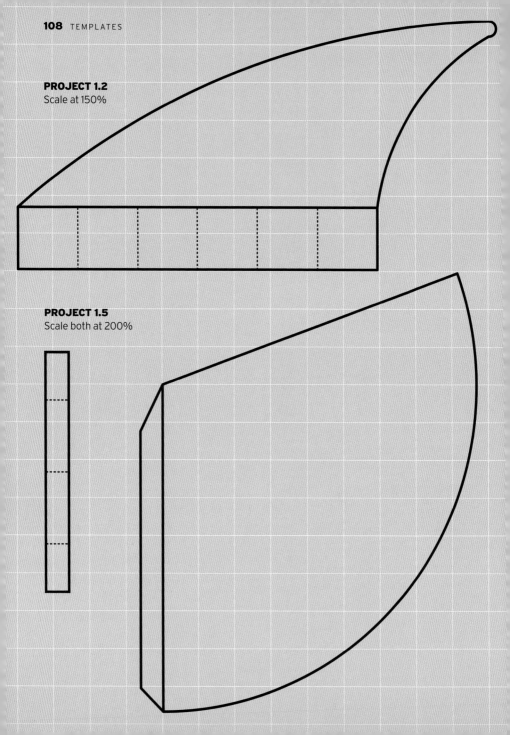

PROJECT 1.2
Scale at 150%

PROJECT 1.5
Scale both at 200%

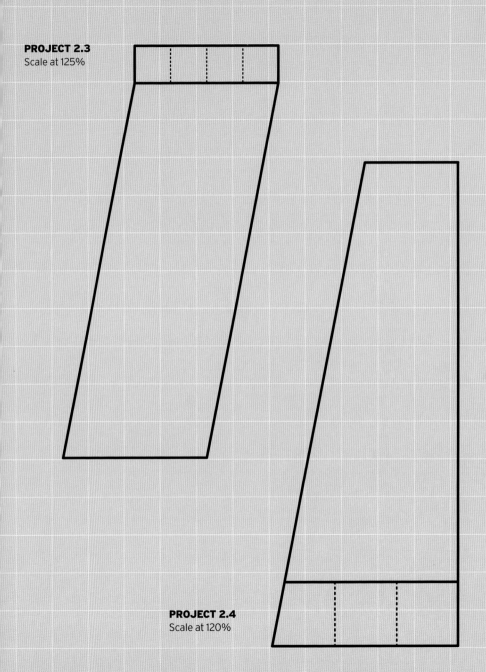

PROJECT 2.3
Scale at 125%

PROJECT 2.4
Scale at 120%

PROJECT 3.1
Scale both at 150%

PROJECT 3.4
Scale at 550%

PROJECT 3.6
Scale at 175%

INDEX

ACKNOWLEDGMENTS

The Ivy Press would like to thank Ian Lambert for his work on the original presentation of this book and Wayne Blades, Paul Jarvis, Ian Lambert, Jyoti Peers, and Lee Suttey for testing the rockets and appearing in the photoshoot.